Barns and Other Out-Buildings

With Information on the Architecture of Farm Buildings

By

D. H. Jacques

British Library Cataloguing-in-Publication Data
A catalogue record for this book is available from the
British Library

Farming

Agriculture, also called farming or husbandry, is the cultivation of animals, plants, or fungi for fibre, biofuel, drugs and other products used to sustain and enhance human life. Agriculture was the key development in the rise of sedentary human civilization, whereby farming of domesticated species created food surpluses that nurtured the development of civilization. It is hence, of extraordinary importance for the development of society, as we know it today. The word *agriculture* is a late Middle English adaptation of Latin *agricultūra*, from *ager*, 'field', and *cultūra*, 'cultivation' or 'growing'. The history of agriculture dates back thousands of years, and its development has been driven and defined by vastly different climates, cultures, and technologies. However all farming generally relies on techniques to expand and maintain the lands that are suitable for raising domesticated species. For plants, this usually requires some form of irrigation, although there are methods of dryland farming. Livestock are raised in a combination of grassland-based and landless systems, in an industry that covers almost one-third of the world's ice- and water-free area.

Agricultural practices such as irrigation, crop rotation, fertilizers, pesticides and the domestication of livestock were developed long ago, but have made great progress in the past century. The history of agriculture has played a major role in human history, as agricultural progress has been a crucial factor in worldwide socio-economic change. Division of labour in agricultural societies made (now) commonplace specializations, rarely seen in hunter-gatherer cultures, which allowed the growth of towns and cities, and the complex societies we call civilizations. When farmers became capable of producing food beyond the needs of their own families, others in their society were freed to devote themselves to projects other than food acquisition. Historians and anthropologists have long argued that the development of agriculture made civilization possible.

In the developed world, industrial agriculture based on large-scale monoculture has become the dominant system of modern farming, although there is growing support for sustainable agriculture, including permaculture and organic agriculture. Until the Industrial Revolution, the vast majority of the human population laboured in agriculture. Pre-industrial agriculture was typically for self-sustenance, in which farmers raised most of their crops for their own consumption, instead of cash crops for trade. A remarkable shift in agricultural practices has occurred over the past two centuries however, in response to new technologies, and the development of world markets. This also has led to technological improvements in agricultural techniques, such as the Haber-Bosch method for synthesizing ammonium nitrate which made the traditional practice of recycling nutrients with crop rotation and animal manure less important.

Modern agronomy, plant breeding, agrochemicals such as pesticides and fertilizers, and technological improvements have sharply increased yields from cultivation, but at the same time have caused widespread ecological damage and negative human health effects. Selective breeding and modern practices in animal husbandry have similarly increased the output of meat, but have raised concerns about animal welfare and the health effects of the antibiotics, growth hormones, and other chemicals commonly used in industrial meat production. Genetically Modified Organisms are an increasing component of agriculture today, although they are banned in several countries. Another controversial issue is 'water management'; an increasingly global issue fostering debate. Significant degradation of land and water resources, including the depletion of aquifers, has been observed in recent decades, and the effects of global warming on agriculture and of agriculture on global warming are still not fully understood.

The agricultural world of today is at a cross roads. Over one third of the worlds workers are employed in agriculture, second only to the services sector, but its future is uncertain. A constantly growing world population is necessitating more and more land being utilised for growth of food stuffs, but also the burgeoning mechanised methods of food cultivation and harvesting means that many farming jobs are becoming redundant. Quite how the sector will respond to these challenges remains to be seen.

BARNS, AND OTHER OUT-BUILDINGS.

There is the barn—and, as of yore,
I can smell the hay from the open door,
And see the busy swallows throng,
And hear the peewee's mournful song.
Oh, ye who daily cross the sill,
Step lightly, for I love it still;
And when you crowd the old barn eaves,
Then think what countless harvest sheaves
Have passed within that scented door,
To gladden eyes that are no more.—*T. B. Read.*

I.—PRELIMINARY REMARKS.

ALL that we need say in introduction to our designs may be embraced in a single paragraph. Let your out-buildings correspond in character with your house, and be as simple in plan and as unpretending in style as adaptation to their uses and an agreeable and appropriate external appearance will permit. A stable should pass for a stable, and not be so elaborate as to be mistaken for a farm-cottage. To build a poultry-house in the form of a palace is equally absurd. Let each seem to be just what it is, and present an example of complete fitness for the purpose of its erection.

Our designs, in general, require very little explanation, and speak for themselves. We present them in the hope that, where they may not be found exactly adapted to particular cases, they may, at least, furnish useful hints toward the thing required. Some of them have stood the test of actual construction and use, and have proved well adapted to their purposes.

II.—LEWIS F. ALLEN'S BARN.

We are indebted to the "Annual Register of Rural Affairs' for the accompanying design. It represents one of the best barns, probably, ever erected in this country, and, although much larger than will generally be required, furnishes a model in most respects for a structure of any desired size. We copy from the "Register" so much of the description as will serve our purpose:

"The body of the main barn is 100 feet long by 50 feet wide, the posts 18 feet high above the sill, making 9 bents. The beams are 14 feet above the sills, which is the height of the inner posts. The position of the floor and bays is readily understood from the plan. The floor, for a *grain* barn, is 14 feet wide, but may be contracted to 12 feet for one exclusively for hay. The area in front of the bays is occupied with a stationary horse-power and with machinery for various farm operations, such as threshing, shelling corn, cutting straw, crushing grain, etc., all of which is driven by bands from drums on the horizontal shaft overhead, which runs across the floor from the horse-power on the other side; this shaft being driven by a cog-wheel on the perpendicular shaft round which the horses travel.

"A passage four feet wide extends between the bays and the stables, which occupy the two wings. This extends up to the top of the bays, down which the hay is thrown for feeding, which renders this work as easy and convenient as possible.

"The floor of the main barn is three feet higher than that of the stables. This will allow a cellar under it, if desired—or a deeper extension of the bays—and it allows storage lofts over the cattle, with sufficient slope of roof. A short flight of steps at the ends of each passage admits easy access from the level of the barn floor.

"The line of mangers is two feet wide. A manure window is placed at every 12 feet. The stalls are double; that is, for two animals each, which are held to their places by a rope and chain, attached to a staple and ring at each corner of the stall. This mode is preferred to securing by stanchions. A pole or scantling, placed over their heads, prevents them from climbing so as to get their feet into the mangers, which they are otherwise very apt to do.

"The sheds, which extend on the three sides of the barn,

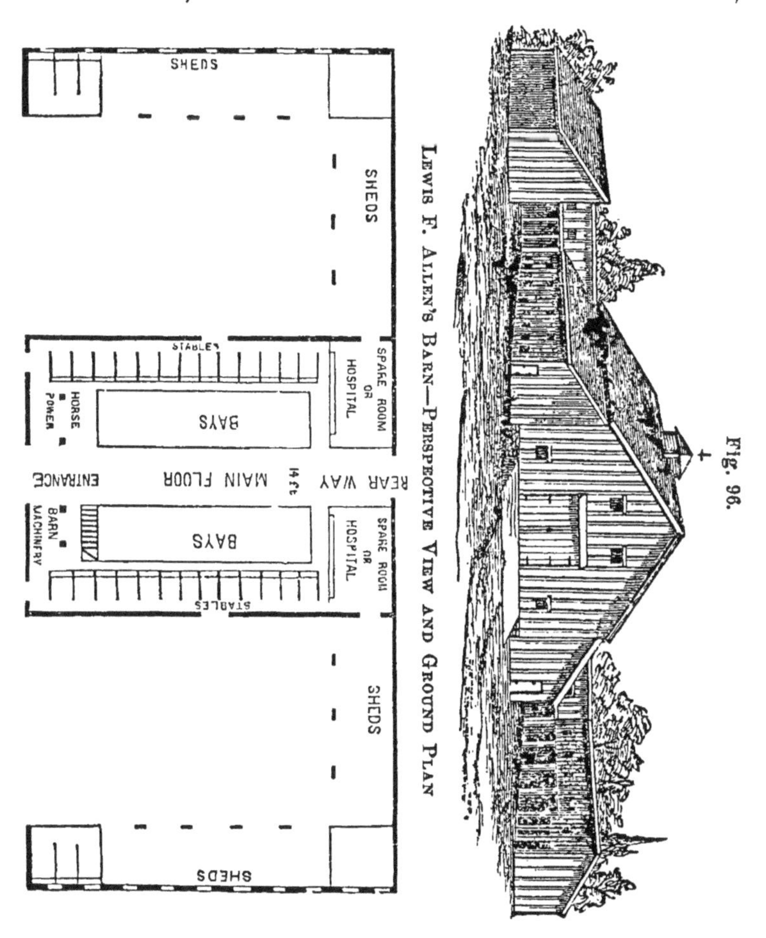

Fig. 96.

LEWIS F. ALLEN'S BARN—PERSPECTIVE VIEW AND GROUND PLAN

and touch it at the rear end, are on a level with the stables. An *inclined plane*, from the main floor through the middle of the back shed, forms a rear egress for wagons and carts, descending three feet from the floor. The two rooms, one on each side of this rear passage, 16 by 34 feet, may be used for housing sick animals, cows about to calve, or any other purpose required. The stables at the front ends of the sheds are convenient for teams of horses or oxen, or they may be fitted for wagon-houses, tool-houses, or other purposes. The rooms, 16 feet square at the inner corners of the sheds, may be used for weak ewes, lambs, or for a bull-stable.

"Racks or mangers may be fitted up in the open sheds for feeding sheep or young cattle, and yards may be built adjoining, on the rear, six or eight in number, into which they may run and be kept separate. Barred partitions may separate the different flocks. Bars may also inclose the opening in front, or they may, if required, be boarded up tight. Step-ladders are placed at convenient intervals, for ascending the shed lofts.

"A granary over the machine-room is entered by a flight of stairs. Poles extending from bay to bay, over the floor, will admit the storage of much additional hay or grain. As straw can not be well kept when exposed to the weather, and is at the same time becoming more valuable as its uses are better understood, we would suggest that the space on these cross poles be reserved for its deposit from the elevator from threshing grain, or until space is made for it in one of the bays.

"A one-sided roof is given to the sheds (instead of a double-sided), to throw all the water on the *outside*, in order to keep the interior of the yards dry. Eave-troughs take the water from the roofs to cisterns. The cisterns, if connected by an underground pipe, may be all drawn from by a single pump if necessary."

III.—MR. CHAMBERLAIN'S OCTAGON BARN.

The accompanying cut represents the ground plan of an octagon barn erected by Mr. Calvin Chamberlain, of Foxcroft. Maine, and described in the "Reports of the Board of Agriculture" of that State.

The plan is on a scale of 15 feet to the inch, which shows the structure to be a trifle over 36 feet in diameter.

"There is a cellar under the whole, eight feet deep, and a

cart-way leading out on a level. The floor is ten feet in the clear; doors same width and height; height below scaffold, seven and a half feet clear; entire height of walls, 19 feet. A door

FIG. 97.

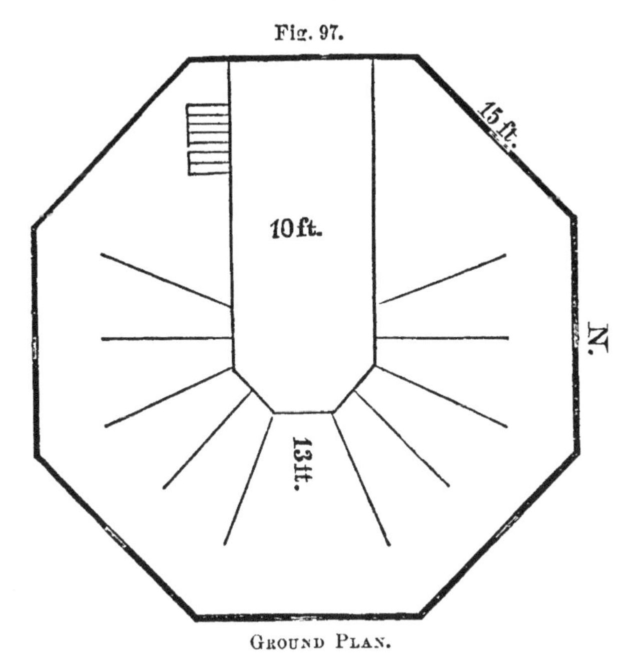

GROUND PLAN.

is shown opening north to the pasture, four feet wide and seven and a half feet high; one south, same size, opening to yard; one on southwest side communicates with other buildings. Stairs lead to cellar and hay-loft. Passage-way behind cattle stalls five feet wide, admitting wheelbarrow to pass at any time to any manure scuttle. Gates hanging to outer wall close passages to stalls, so that any animal may occupy its place untied. Side-lights at large doors, and a large window on opposite side, one sash of which slides horizontally, light the stable. Four large windows, set quite up to the plates, light the hay-loft. These let down at top, and are left down half the year; the two-feet projection of the roof protects them from all storms. Cellar is lighted by four double windows and the side-light at head of stairs. The open space, 13 feet long, at end of floor, admits the horse, so that the hay-cart is brought to the center of the barn for unloading. A scaffold 13 feet long is put over the floor, and 12 feet above it."

This small barn, Mr. Chamberlain says, will store 20 tons of hay.

IV.—MR. BECKWITH'S OCTAGON BARN

The annexed cut represents the basement plan of the barn erected by E. W. Beckwith, Principal of the Boys' Boarding School, at Cromwell, Middlesex County, Connecticut, in September, 1858.

The beauty and convenience of the arrangement for stalls and feeding can be seen at a glance. The octagon form is adopted because it is best adapted to inclose the desired plan.

This building, 30¼ feet short diameter, 12½ feet each side, or 100 feet inside circumference, and 13 feet each outside, or 104

Fig. 98.

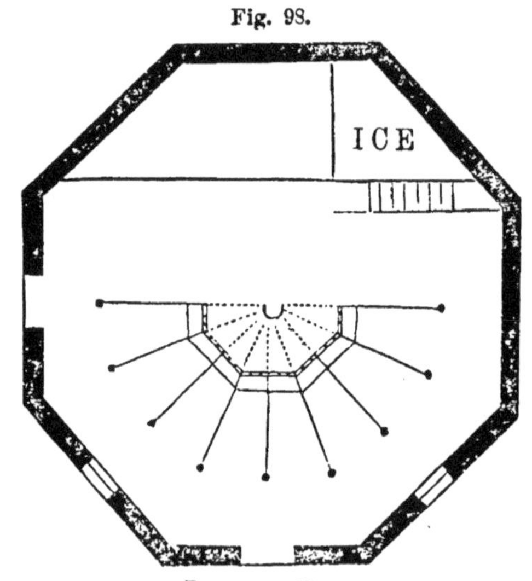

BASEMENT PLAN.

feet circumference when the wall is 14 inches thick, as in the present case, incloses an area of 750 feet.

The wall is grouted stone work, laid up between planks cut the right length for each inside and outside of angle, held to the proper distance apart by cast-iron clamps pierced with holes at each end to receive the iron dowels driven into each edge of the planks. These planks, when in an upright position on the wall, should be plumbed and staylathed preparatory to laying the stone. The basement floor is cemented, the horses standing on a movable slat-work, which keeps the bedding dry. The height of this story should be eight feet; the clear space from the stalls to the wall, four feet wide; the stalls six feet long, including manger-box, which leaves a circle in the center about ten feet diameter as the base of a cone, over which all the feed is thrown down to the animals. Under the cone is a fine place for a water-tank or pump.

The remaining space, when not wanted for stalls, furnishes room for cleaning off horses, for storing roots, for an ice-house, or any other purpose for which it may be wanted.

The feeding place is a hole about three feet square over the apex of the cone, which can be covered with a scuttle.

The walls are 26 feet high from the foundation, giving 16 feet altitude above the barn floor, which can be left clear and open to the roof, thus allowing the hay to be deposited in any direction and to any required proportion of the space; a gang-way to the feed-hole being left, or cut afterward, at option. There is one door, 9 by 10 feet, to this floor, for carriages, etc., the hay being taken in at a window on the up-hill side. Of course a place would be partitioned off if carriages are to be housed in the barn.

The cost of this stone barn, covered with mastic roofing at five cents a foot, will be about $325.

The walls cost $230, but closer personal attention would have made them cheaper. A wood barn on the same base-ment would have cost at least $40 more, and not be as good for many reasons.

There is nothing to burn by fire but one floor, and the roof and the walls would be left for another.

The utility of narrow stalls, in this case five feet wide at the broad end and two feet at the manger, may be questioned by some; but you have that matter entirely according to fancy, the peculiar feature of this plan being that they all point to the center. It is peculiarly adapted to those gentlemen who wish to keep horses and cows, and be able to feed them without too much labor or time and exposure to dirt.

You can have a hired man or not, as you choose, which is sometimes desirable. This plan, if not adopted by others, may serve a good purpose as a suggester.

V.—A CIRCULAR BARN.

The barn, plans of which are herewith presented, was built by the Shakers of Berkshire County, Massachusetts, and is certainly worthy of the attention of farmers contemplating the

Fig. 99.

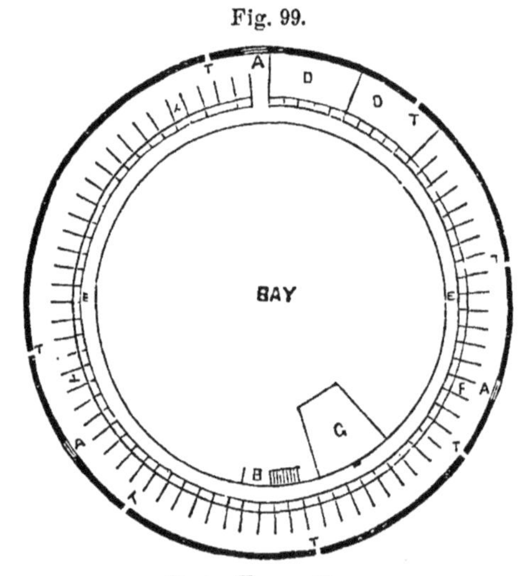

FIRST FLOOR PLAN.

A., doors;* B., stairs; D., calf-pens; E., alleys; F., stalls; G., granary; H. double doors; T., windows.

erection of barns on a large scale. It is 100 feet diameter, built of stone—a material that is very abundant in that part of Massachusetts—two stories high, the first one being only seven and

* An error in the plans represents the doors as windows, and *vice versa*.

a half feet between floors, and contains stalls for seventy head of cattle, and two calf-stables. These stalls are situated in a circle next the outer wall, with the heads of the animals point-

Fig. 100.

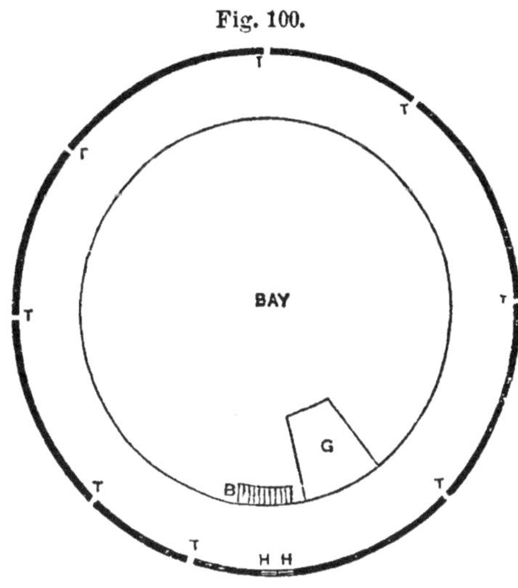

SECOND FLOOR PLAN.

ing inward, looking into an alley in which the feeder passes around in front of and looking into the face of every animal. The circle forming the stable and alley-way is 14 feet wide, inside of which is the great bay. Over the stable and alley is the threshing-floor, which is 14 feet wide and about 300 feet long on the outer side, into which a dozen loads of hay may be hauled, and all be unloaded at the same time into the bay in the center.

There should be a large chimney formed of timbers open in the center of such a mass of hay, connecting with air-tubes under the stable floor, extending out to the outside of the building, and with a large ventilator in the peak of the roof. We should also recommend an extension of the eaves beyond the outer wall, by means of brackets, so as to form a shed over the doors, and the manure thrown out of the stables and piled against the wall.

VI.—A SIDE-HILL BARN.

We copy the accompanying plans and the description from the *American Agriculturist* for September, 1858, where a per-spective view of the barn is also given.

Entering the barn at either end, as shown in the main floor plan, there is a floor, either 12 or 14 feet wide, as may be most convenient, which passes through the entire length. On

Fig. 101.

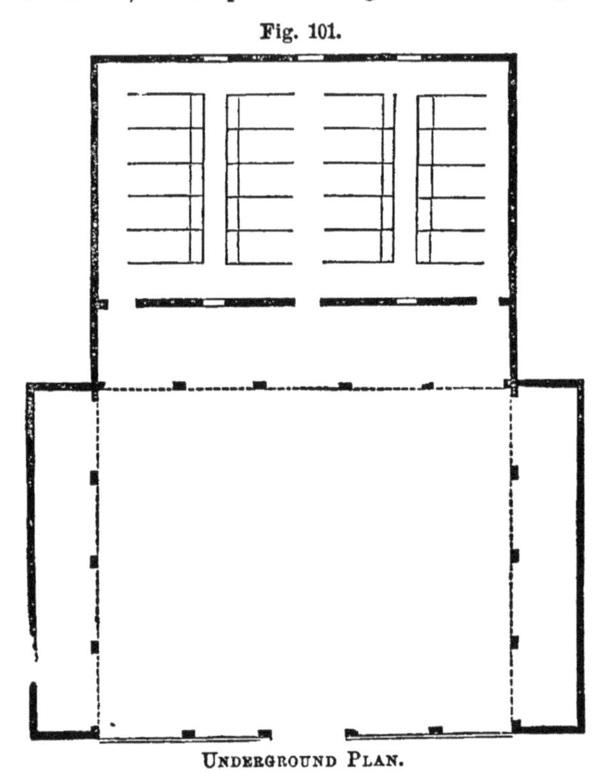

UNDERGROUND PLAN.

one side is a large bay for hay or grain in the sheaf. Oppo-site, in part, is another bay. Next to that a passage of five feet wide, to carry out straw or hay to throw down below into the yard. Next to the passage is a granary, and adjoining it a tool-house, or area for threshing machines, straw-cutters, etc., with a partition off from the floor, or not, at pleasure. Nine feet above the floor, on each side, should be a line of girts, connecting the inner posts, on which may be thrown loose poles to hold a temporary scaffold for the storage of hay, or grain in the sheaf, when required. By such arrangement the barn can be filled to the peak or ridge-pole, and the ventilator above will carry out all the heated air and moisture given off from the forage stored within. Slatted windows, or side ventilators, may be put in the side next to the yard, if required. The roof has a "third" pitch, or one foot rise to two feet in width, which lasts longer and gives more storage than a flatter one.

The frame of the barn above is 60 by 50 feet, with posts set upon stones below, to support the overshot sill, as shown in

Fig. 102.

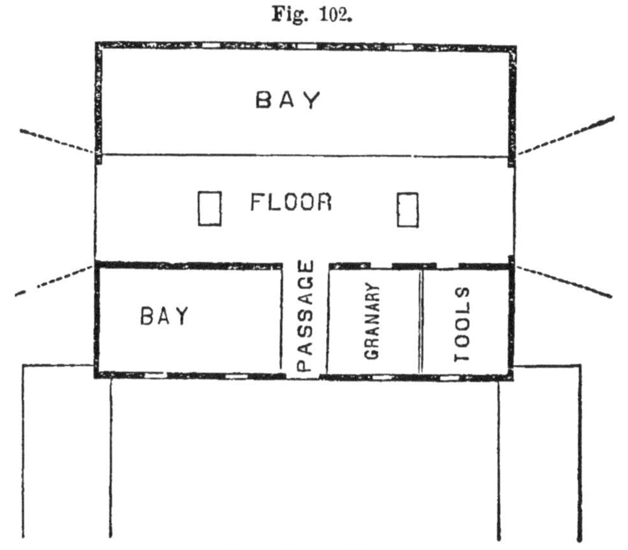

Main Floor Plan.

the ground plan. Underneath are four lines of stalls, two on each side of the center passage-way, heading each other, with a four-foot feeding alley between them, receiving the forage from above, from which it is thrown into the mangers, two and a half feet wide, to which the cattle are tied or chained. The stalls are double, allowing two animals, if neat stock, in each. They are tied at the sides next the partitions, to prevent injury to each other. On the hill-side are three windows in the upper part of the wall, to admit light and ventilation, either glazed or grated, as may be necessary.

The advantages of a side-hill barn are, the warmth of its stables in winter and their coolness in summer; storage for roots, if required; much additional room under the same roof, but not, we think, at diminished expense; and greater compactness of storage than in one on the common plan.

But it is essential to the comfort and convenience of the side-hill barn that it be well embanked with earth, so that the falling water may freely pass away from the walls; and that the stables and yards be well drained. Without these precautions, such barns are little better than nuisances, the rains and melting snows flooding everything beneath the building, and in the yards and sheds below.

There should be a flight of stairs (not represented in the plan) from the underground floor to that above.

Shelter Cheaper than Fodder.—An improvement on our present practice of shelter, and care of our animals, would be an equivalent to an actual shortening of winter. It can hardly be questioned that exposure of cattle to extreme cold injures their health, and thus interferes with the owner's profit. Chemical physiology teaches us that warmth is equivalent to a certain portion of food, and that an animal exposed to more cold will eat more, and one better housed and warmer kept will eat less. To keep an animal comfortable, therefore, is to save food; and this alone is a sufficient inducement to provide that comfort to the full extent.*—*Maine Agricultural Report.*

——Every animal should have its own particular stall in the stable, and should be allowed in no other.

* It is asserted, on good authority, that exposed animals will consume a third more food, and come out in the spring in worse condition.

VII.—STABLES.

The subject of stables—their construction, arrangement of accommodations, etc.—is one to which a volume might profitably be devoted; but our present object is merely to furnish a

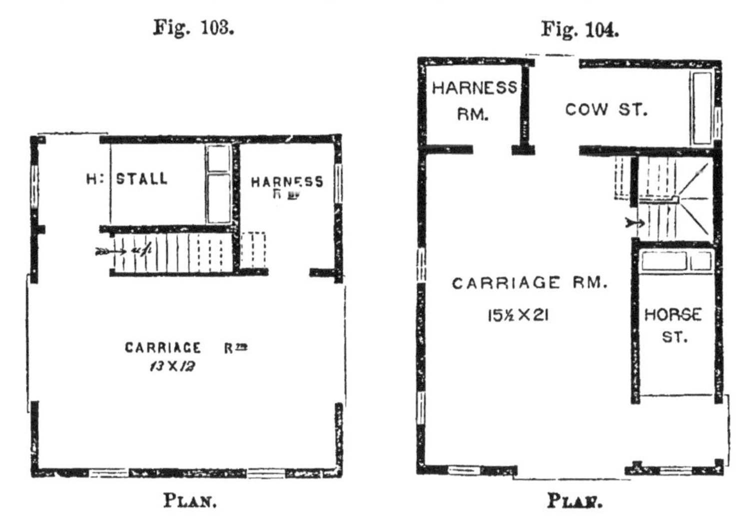

Fig. 103. PLAN. Fig. 104. PLAN.

few designs adapted to execution in connection with country houses and villas, and to show how they may be planned, as

Fig. 105.

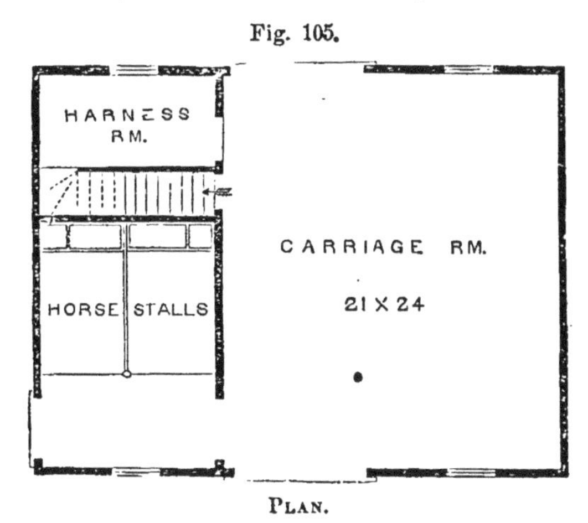

PLAN.

in fig. 103, for one horse and carriage; in fig. 104, for one horse and two vehicles; or, as in fig. 105, with which we give an elevation (fig. 106), for two horses and three vehicles.

Fig. 106.

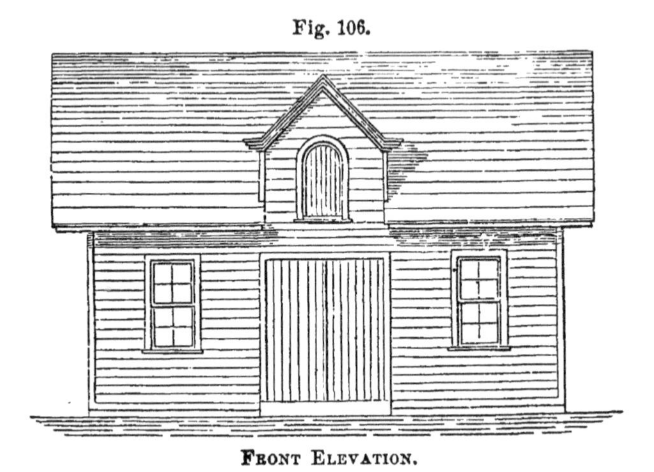

FRONT ELEVATION.

Constructed of wood in a proper manner, fig. 103 will cost $125; fig. 104, $185; and fig. 105, $275. Built of brick, they will generally cost a little over a third more.

ELEVATORS IN BARNS.—In large barns the pitching up of the hay into the upper part of the bays is a very laborious process and requires considerable time. In such cases an *elevator*, like that of the best threshing machine, to be worked by the two horses removed from the loaded wagon of hay, may be profitably employed, greatly lessening the labor and quickening the operation. The same elevator would be used in carrying threshed straw from the machine to the bays. The simplest and best elevator for this purpose is made of a light, inclined board platform, four feet wide, on each side of which a rope or endless chain runs, connected by cross-bars, a foot or two apart, which slide over the upper surface of this platform, and sweep the hay upward as fast as pitched upon it.

VIII.—AN OCTAGON POULTRY HOUSE.

This design is selected from Bement's "Poulterer's Companion." It has been executed, we believe, near Factoryville, Staten Island. It is ten feet in diameter and six feet and a half high. The sills are 4 by 4, and the plates 3 by 4 joists, halved and nailed at the joints. It is sided with inch and a quarter spruce plank, tongued and grooved. No upright timbers are used. The floor and roofing are of the same kind of

Fig. 107.

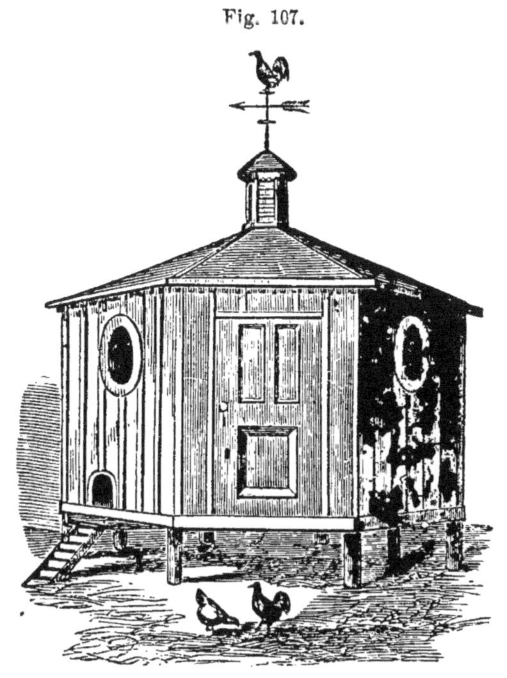

PERSPECTIVE VIEW

plank. To guard against leakage by shrinking, the joints may be battened with lath or strips of thin boards. An eight-square frame supports the top of the rafters, leaving an opening of ten inches in diameter, on which is placed an octagon chimney for a ventilator, which makes a very pretty finish. The piers should be either cedar, chestnut, or locust, two feet high, and set on flat stones.

The letter D designates the door; W, W, windows; L, latticed window to admit air, with a shutter to exclude it, when necessary; E, entrance for the fowls, with a sliding door; P, platform for the fowls to alight on when going in; R, R, roosts placed spirally, one end attached to a post near the center of the room, and the other end to the wall; the first, or lowermost one, two feet from the floor, and the others 18 inches apart, and rising gradually to the top, six feet from the floor. These roosts will accommodate 40 ordinary-sized fowls. F, F, is a board floor, on an angle of about 45 degrees, to catch and carry down the droppings of the fowls. This arrangement renders it much more convenient in cleaning out the manure, which should be frequently done.

Fig. 108.

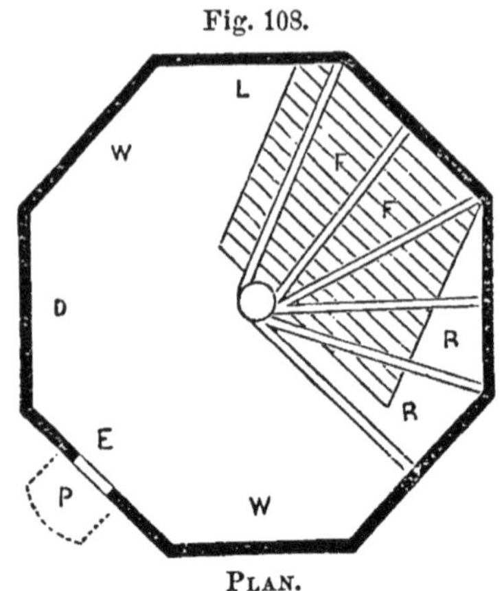

PLAN.

The space beneath this floor is appropriated to nests, 12 in number, 15 inches wide, 18 inches deep, and 18 inches high. In order to give an appearance of secrecy, which it is well known the hen is so partial to, the front is latticed with strips of lath. By this arrangement a free circulation of air is admitted, which adds much to the comfort of the hens while sitting.

The object of placing this house on piles is to prevent the encroachments of rats, mice, skunks, etc., and is a good method, as rats are very annoying, especially where they have a good harbor under the house, often destroying the eggs and killing the young chickens.

TWO ERRORS.—It is an error to build a house upon a side-hill with an "underground kitchen;" but it is a greater error to build a barn without such a room upon the down-hill side.

and if possible having a southern exposure. In this room all the horned cattle should be stabled, having a yard to themselves entirely separate from any other stock. The horse stable should always be on the ground floor, with an entrance from a separate yard.

IX.—AN OCTAGON PIGGERY.

The accompanying design shows the plan of an economically constructed and convenient piggery. It may, of course, be enlarged to any desired extent without any change of form or

Fig. 109.

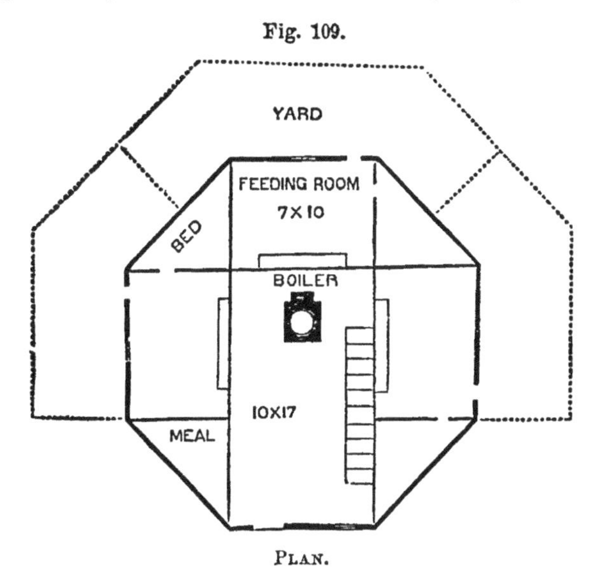

PLAN.

arrangement. The elevation may be similar to that of the poultry-house (fig. 107), and should have sufficient height to furnish a good upper room for storing corn, etc., for the swine.

X.—AN ASHERY AND SMOKE-HOUSE.

An ashery and smoke-house combined may be economically built as represented in our design. The first story, or ash-pit, should be built of stone or hard brick, and be provided with an iron door. The walls need not be more than from six to eight feet in height. The ceiling should be lathed and plastered.

The smoke-house story above may be of wood. It is entered in the rear on a level with the ground. Four tin tubes, introduced through the floor, admit the smoke from the ash-room below, where the fire is kindled. This arrangement precludes all danger from fire, secures the meat against being overheated

Fig. 110.

AN ASHERY AND SMOKE-HOUSE.

in smoking, and gives a clean and convenient smoke-room. It may be ventilated either through the gable or the roof.

A side hill situation is by no means essential in this mode of construction. Both stories may be above ground, the smoke-house door being reached by outside stairs or a step-ladder.

XI.—AN ICE-HOUSE.

The first grand essential in the construction of an ice-house is the perfect inclosure of the space to be occupied by the ice with walls and floors which shall prove non-conductors of

Fig. 111.

A CIRCULAR ICE-HOUSE—PERSPECTIVE VIEW.

heat. The second important point is to secure perfect drainage. These conditions attained, the rest is comparatively unimportant.

A common and entirely effective mode of constructing an ice-house is thus described:

The frame or sides should be formed of two ranges of upright joists about six by four inches; the lower ends to be put in the ground without any sill; the upper to be morticed into the timbers which are to support the upper floor. The joists in the two ranges should be each opposite another. They should then be lined or faced with rough boarding, which need not be very tight. These boards should be nailed to those edges of the joists nearest each other, so that one range of joists shall be outside the building and the other inside the ice-room, as shown in fig. 112. Cut out or leave out a space for a door of suitable dimensions on the north or west side, higher than the ice will come, and board up the inner side of this opening so as to form a door-casing on each side. Two doors should be attached to this opening —one on the inner side and one on the outward, both opening outward. The space between these partitions should be filled with charcoal-dust, tan, or saw-dust, whichever can be the most readily obtained.

Fig. 112.

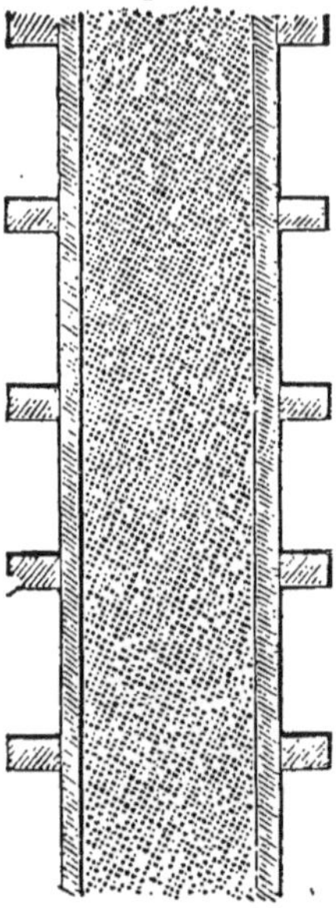

The bottom of the ice vault should be filled about a foot deep with small blocks of wood or round stones; these are leveled and covered with wood-shavings, over which a plank floor to receive the ice should be laid; some spread straw a foot thick over the floor, and lay the ice on that. A floor should also be laid on the beams above the vault, on which place several inches of tan or saw-dust. The roof should be perfectly tight, and it is usually best to give it a considerable pitch. The space between the roof and the flooring beneath should be ventilated by means of a door or lattice window in each gable. The drain can be constructed in accordance with the situation, the only things requiring attention being to have it carry off all the water settling at the bottom, and not be so open as to allow the passage of air into the vault.

Fig. 113 represents a section of such an ice-house. We give a perspective view of a circular ice-house, which is constructed on the same principle. It may advantageously be executed in

Fig. 113.

concrete. Ventilation is secured by leaving a small aperture in the peak of the roof, protected by a hood or cap, as shown.

Should an underground house be preferred, the plan of building can be the same; or a less expensive method may be used. A side-hill having a northern exposure affords a desirable location. In such case one end of the house is usually above ground. The boards can be of the cheapest description, and the space or air-chamber filled in with straw; the ground forming the support to the whole. No less attention should be paid to draining than in the other case; and when in use, the space between the ice and the peak of the roof should be filled with straw.

The House.

XII.—AN APIARY.

Fig. 114 represents a design for a rustic apiary or bee-house, which strikes us as being far more beautiful and appropriate than the elaborately ornamented temple or palace-like structures we sometimes see The mode of its construction is readily

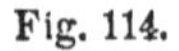

Fig. 114.

Perspective View.

seen. It may, of course, be made of any desirable size on the same plan. [For directions in reference to the construction of hives, the best site for an apiary, and instructions in bee-keeping, see " The Barn-Yard."*]

---

* The Barn-Yard: a Manual of Horse, Cattle, and Sheep Husbandry; or, How to Breed, Rear, and Use all the Common Domestic Animals. Embracing Descriptions of the various Breeds of Horses, Cattle, Sheep, Swine, Poultry, etc.; the "Points" or Characteristics by which to Judge Animals; Feeding and General Management of Stock; How to Improve Breeds; How to Cure Sick Animals, etc. With a Chapter on Bees. Handsomely illustrated. Now published with Garden and Farm. $1.75.

How many expensive, not to say fatal, errors in the buying, selling, breeding, and management of farm-stock might be avoided by means of the practical information and plain common-sense advice condensed into this comprehensive and thorough little Hand b ok!

XIII.—A PLAY-HOUSE.

Build your children a play-house of some sort. A very rude affair will please them, but something similar to the accompa-

Fig. 115.

Perspective View.

nying design will please you too, and be a highly ornamental feature in your grounds. The construction is simple, but the effect is verv fine.

Materials for Rustic Structures.—In order to succeed in constructing rustic work, the first thing is to procure *the materials.* All such objects as may be exposed to the weather should be of the most durable wood, of which red cedar is best. For certain purposes, white oak will answer well, but as it is essential to have the bark remain on, the wood should be cut at a time of year when this will not peel or separate. If cut toward the close of summer, the wood will last about twice as long as when cut in winter or spring. A horse-load or two of boughs or branches of trees, of which a goodly portion may be curved and twisted, from one to six inches in diameter, will constitute the materials for a good beginning.—*J. J. Thomas.*

XIV.—A RUSTIC GARDEN HOUSE.

A rustic structure, like the one here represented, when covered with vines and climbing shrubs, forms one of the most

Fig. 116.

Perspective View.

beautiful and appropriate objects that a lawn or flower garden can boast. Furnished with rustic seats, it becomes an attractive summer resort in which to work or read.

Milton Keynes UK
Ingram Content Group UK Ltd.
UKHW040946130224
437765UK00001B/48

9 781446 531341